Forward for Alternate Energy

The purpose of writing this story is two fold. First is to emphasize the seriousness of Global Warming. The only way to solve it is a GLOBAL EFFORT. There can be no "Sitting on the sidelines". All Countries on our Planet must be involved! The defeat of the "Big 3"; which are Oil, Insurance, and Internal Combustion Cars, An alternative to these must be found. The people from the "Big 3" can use their tremendous resources, and be placed in these alternate fields. This can be done; and made practical, but only by a GLOBAL EFFORT". The Insurance which is used for all aspects of life, can be neutralized from the uselessness and greed factor and made into a more useful entity.

Second, is romance, and young love in the story. The total devotion To a cause or a person has always been a good glue for a story.

Earl Hammers

San Jose, California 1-8-2015

Alternate Energy Inc.

The day began like most for Dave Jones. A day of running a big

Company; and trying to get others to "see the light". This was sharing

Their knowledge with others. His wife Clarisa came in with a cup of

Coffee. She was a woman in her mid 40's in age. She had curly brown

Hair, with one Green Eye and one Blue Eye. She had short eye lashes

and a trim figure. She was wearing a Robe and slippers, and

no makeup (Itwas7 in the morning!) Her hair was nicely combed and

She had a diamond ring on the third finger of her left hand. She had

no other jewelry. She had a cup of coffee which she presented to Dave.

Dave Jones was a well built and vigorous man, in his mid 50's. He

had short Brownish Gray hair. He had one Gray Eye and one Brown Eye. He

was clean shaven, no mustache or Beard. He had a diamond ring also on the

Third finger of his left hand. He took the coffee.

Clairisa said, "Dear, you seem somewhat anguished, and troubled is

there anything wrong at the office or here? "

"Yes dear, I am happy with the office, but, I think about others that

are suffering the Same fate that we had back in the 20'Th century",

said Dave.

"The 20'th century!! But that was 300 years ago; surly we don't have

troubles like That, with all of our improvements ", said Clarisa.

"Not for us, but for a planet that we have been asked to visit and

help if possible. A planet much like we used to be." Said Dave.

"That's wonderful news dear. Just think to represent our world, to a group of people We have never met. I'm excited ", said Clarisa.

"Yes. It is a tremendous opportunity to do what I have wanted to do, but, I just hope I can do some good. I was informed of this just last week." said Dave.

Last week!! Do you have a group together to go? Who will you take?" said Clarisa.

Yes. I am putting a team together now. It will be our top people, plus The crew and Support staff from the Planet Alliance. The crew of the Ship to get us there as well", saidDave.

"Wow! The Planet Alliance, they handle all of the exploration and Research. That's a Great honor to have them here on this endeavor. ", Said Clairisa. She threw her arms around Dave and gave him a full kiss.

"Yes. I just hope it goes well. They have tremendous problems on this Planet. We are Supposed to go to the ship's captain today, to make the final arrangements." said Dave.

"Today!! Good heavens, how long will you be gone? There was something I wanted to talk over with you ", said Clarisa.

"We will be gone at least 5 years, maybe more. What was it you wanted to talk over?" said Dave.

"I know you are very busy…. Oh never mind, we will talk about it when you get back. By the way you better get on your horse, if you want to make your meeting", said Clarisa.

"Thank you dear for being so understanding." He but his arms around her and gave her a full kiss on the lips. "I do not know what I would do without your faithfulness. I…..",said Dave. He quickly got

dressed, kissed his wife again, and then departed for the Earth's
Surface. They were in the Cloud City, which was where they lived. He
got in his air vehicle and left for Earth's surface. The Cloud city had
been the crowning achievement of Earth's building technology. There
Were several of them scattered about the planet.

The people who were appointed by Dave Jones, and to a lesser
extent by Robert Paul, And his wife Betty were the following: James
Jameson, John Johnson, Ray Robinson, and Mike Michelson, were
carefully hand-picked by Dave. They had tremendous achievements
And respect from their colleagues. This was in college, and in the
Business world on Earth. Dave knew them by reputation and
Performance, both in business, and in the world.

Chapter 2

When Dave arrived at the office he was met by his board of
Directors, plus James Jameson and Robert Paul. Robert Paul was his
Second in command, and James Jameson in charge of production.

"Well Dave, said Paul Are you ready to go into the lion's den over
there at the Space Agency, the Planet Alliance? I hear that they are a
Real tough lot, especially on those that are not in the service, or space
I want you to know that the board is behind you all the way".

"Thanks Rob. I appreciate that. I'm going to need all the help I can
get. What have you heard About the Captain…., Kirk Christopher I
believe? I know he runs a taut ship, but I believe he is reasonable

I actually look forward to meeting him ", said Dave.

The rest of the board said in unison" We wish you good luck in this endeavor. May the wind be at your back! May the people of Earth be proud of you! "

"Thanks a lot Ladies and Gentlemen I don't know what to say. I…"

"Forget it pal! Said Robert. He extended his hand and gave it a shake. I'm going with you, Betty too", said Robert.

"Thanks a lot I'll need all the moral support I can get. Where is Betty anyway? ", said Dave.

"Here I am. I'm ready to go to the space agency right now", said Betty. So with the applause of the board, they left for the Space Agency, they got there quickly because it was a short trip by Pod, which was the main order of travel in the city.

Robert Paul was a friend of Dave's for many years. They had Served in the Planet Alliance As young Engineers. They had enjoyed, and liked traveling to other worlds, but didn't like all the saluting and Other military crap you had to put up with. They found the company's Too static, and unwilling to share their knowledge with other worlds, The point was partly justified. However in the recent time the agency had contacted him about a planet that had requested help, and since they were a member, they decided to comply. This in fact, was the purpose of the meeting. In fact it was a similar situation that he had meet his wife. She had been an Engineering advisor on a similar Project. It had been love almost at first sight.

Dave had meet his wife through a New Year's Eve party after he was in the service. Robert Paul had played match-maker on this episode.

When they got to the agency, they were met by Kirk Christopher, Christopher Jones, and Barbara Jones, plus others from the agency. Kirk Christopher said, "Let's open this meeting. Our ship is ready. My X.O. and chief Engineer here Can give their approval".

Christopher Jones spoke, "Our crew has made the final mission Preparations, and can be ready to get underway in a few days."

Barbara spoke, "The ship is in top notch shape, and we can be ready to give you a pleasant Voyage. We realize the urgency of the matter."

Dave said, "Great. I can have my team ready to go in a week. There are 30 in total, not Including Robert Paul, Betty and myself."

"Good. Said Kirk We can be ready in a week without further delay" They shook hands all around, and departed the agency.

"Wow! Said Betty, after they left "They don't mess around. I'm glad that we can help another world. To make a difference is very important, and satisfying"

Robert Paul said, "Did you think it would be that quick Dave?"

"No. , said Dave. I expected a lot of crap, and static from the Captain. I'm glad that he saw the urgency of the matter". They returned to their Homes, and Dave was prepared to talk with Clairisa about "The Matter" she spoke of.

"I'm thrilled for you dear. She put her arms around him and gave him a big kiss. She resolved herself to the trip, and "the Matter" she wanted to discuss would be put on hold.

The week went by rapidly, and when Dave and Robert were to leave

The good-by's were short and sweet. Clarisa said, all the best to you on

Your voyage, and your mission. Bring that mug of yours home, there is

something I want to discuss with you." She gave him a long kiss,

And left. Robert and Betty said good-by to Clarisa and went aboard...

The other people from Alternate Energy arrived, and put their gear

aboard. They consisted of Scientists, Engineers, and Technicians. They

came aboard, and moved to their quarters. The crew of

Explorer 2 was already aboard since they were in the service. The

Explorer 2 was one of the smaller vessels in the fleet, but were

Outfitted for scientific research. The journey was uneventful, and due

to high light speed, and other factors only took several months.

They arrived at Rowpedia in the Alpha-Omega Star Group .

################

James Jameson head of manufacturing and production, who was

to speak upon, arrival. He was meet by Cintheianon who was head of

the delegation for the group, Crusade for a Better World. The Captain

of the ship, Kirk Christopher along with his Science Officer Rebecca

James. Captain Kirk Christopher did the introduction……

"I am Kirk Christopher; I would like you to meet James Jameson

Head of Manufacturing and Production for Alternate Energy".

James Jameson said," A pleasure to meet you at long last. I read the

Briefing about you, and It is a big task you are taking on." He extended his hand to Cintheianon as he said this.

Cintheianon extended her hand as well and said, "A pleasure to meet you, and we the people of Rowpedia thank you".

Kirk introduced Dave Jones the President of Alternate Energy. Dave Jones said," A pleasure. I've yearned for the opportunity such as this; I'm more glad perhaps then you."

Kirk introduced the rest of group from Earth they were Robert Paul, his wife Betty, along With the other Engineers and Technicians involved with the project.

Cintheianon said," This is my second in command Angelanon. She will work closely with your People, so we can coordinate our efforts."

Angelanon said," a pleasure ladies and gentlemen. Thank We would like you to stay in orbit around our planet for a time for logistical, and Defense until we can get the ball rolling." The Captain made his Acknowledgement Of this agreement, and returned to the ship.

Cintheianon addressed the group who were gathered in one open Space "Ladies and gentlemen we have come to grave crisis, and we need your help to convince our leaders and our people of the need to change our life style. We are up against a powerful group that will stop at nothing to achieve their ends. These are: the Big Three Car Company's Unbeatable, Dominate, and Absolute. The three big Oil Company's Unconquerable, Omnipotent, and Unbeatable. The final one And perhaps the worst of the Big Three Insurance company's

Unbeatable, Dominate, and Necessity. The names tell it all. they

are ruthless, predatory, and will do whatever it takes to achieve their

Ends, to each other and our world. They have complete control over

many aspects of our world. They have, as a result of their activities,

reached a point of massive climate change that has been a Catastrophic

Effect on our world. We have a Government Congress both House and

Senate totally under their control by bribes, money, and other factors.

It is through this massive infusion of cash that can finance a relentless

Campaign against our group. Will you help us!!? "

   The group responded in unison" We the people of Earth, and with

The help of other worlds in the Planet Alliance can bring tremendous

pressure to bear on the perpetrators of this crisis. We don't interfere

usually with the affairs of other planets, but, this is a different situation.

 We will be glad to help you in any way we can."

      Cintheianon said," Good. Let's get settled, and get to work. My

Assistant Angelanon   Will get you to your quarters."

      So the process began of teleporting down all gear, supply's, and

Personnel. This took a few hours, and the work could begin in

earnest....

                     ****

Meanwhile back on Explorer 2 a meeting was underway.

To speak was X.O. Chris Jones, "Well Chris what do you think of this

Mission these people are going to take?'

      Chris spoke," Well sir, It's a big task they have taken on. I hope they are up to it."

Barbara spoke, "I believe with my team's expertise and the people of Earth, as well as the People below us are up to the task sir."

Chris spoke," They have pretty good technology, and with our help they will succeed."

Barbara spoke," I believe with these people's dedication, and determination to succeed I believe they will win the day. I have never seen such a determined, and a singleness of purpose."

Captain Kirk nodded in acknowledgement, "I agree gentlemen, and Lady. We will remain and be of assistance when needed. Very well, Carry on, and dismissed'. The meeting ended.

Chapter 3

Cintheianon and Angelanon had similar backgrounds. They both were children, ironically of big company executives. Cinthianon was Dave Davis's daughter of Omnipotent Oil. Angelanon was Mary Martin's daughter of the insurance giant Unbeatable. They were a Continual thorn in the side of their parents, and embarrassment too. They went to separate schools, but met in college. Their similar Interests, help forge their work, and friendship. They caused some Trouble for the companies, but nothing to make a real dent in them. They formed the group Crusade For A Better World, but this too had little effect. Their parents were always pushing suitors at them, but, they rejected them. The reason for this was, they were immature, plastic, and stogies. They turned in desperation to the Planet Alliance, in which Rowpedia was a member. They had a promise of help. they had reason for hope to end the crisis.

The climate on Rowpedia was as follows: It was generally a very Temperate climate. They were the fourth planet in their system. They had Deserts, and Poles. They had 3 oceans, and several seas. They had several Mountain Ranges which averaged 15 to 20 thousand meters. The peaks 100 to 200 meters above this. They had seven continents, 50 Islands and hundreds of Lakes sprinkled about on the Planet's surface. All and all a nice planet. However, in the past 50-60 years the climate had become extreme. The temperatures in many places were well over 110 f. The other places were below freezing. The rain, in other area's were extreme as well, sometimes 8-10 inches in a day. This was always accompanied by hurricane force winds of 150 M.P.H. !!
The areas of Extreme draught where the temperature was hot, no rain. This was the summer patter.
In the Winter it was the following: There were extremely cold temperatures such -50f and more, this was accompanied by 70-90 M.P.H. winds and no precept. In the other, snowfall of several feet at a time accompanied by winds of 70-90 M.P.H. To add to the mix, there were extremes here as well. One was cold of -50C. The others were boiling hot in the same season.

Despite this the People struggled to survive. They were thousands that died, from starvation, or cold. But, despite this there were areas of relatively normal temperatures and climate. The lucky few, and the Perpetrators of the crisis were here.

****

The first few days were more or less a get acquitted time. The people of Earth, had featured Leaders such as: John Johnson head of United Pod Systems, Ray Robinson head of Full Power Solar, and Mike Michelson head of Unbeatable Electric. They had support staff as well, which consisted of Engineers, and Technicians and other workers. The total delegation was about 100. They were not all present around the Table. Cintheianon and Angelanon were there as well. The heads of the Various divisions of the Crusade for a Better World were sitting in chairs in the same room as the table. The table was round, and was elevated above the room on a dais.

Cintheianon brought the meeting to order by saying," I will introduce key members of the Crusade, they are: Robert Jenningsnon Head of Freedom Solar, Pat Smithnon head of Abundant Wind, Joe Jonesnon head of Eternal Tides Inc. and Ruth Gundersonnon head of Dependable Geothermal. These C.O.'s along with their assistants will guide the people of Earth to get the full picture of what's going on here on Rowpedia. You will be here to advise us since you have already dealt with this problem".

Pat Smithnon spoke first," We have many places here on Rowpedia that our windmills, and the ones you send us can be effective.

One area that we would like you to review, and consider is the desert. There is a great deal of wind there. This is due to the heat differential between there and the coast."

Robert Jenningsnon spoke" The desert is also an ideal place for our Cells to be placed, due to the space there. We would like to advise as well, and other places you might find suitable. The farms can be several Kilometers in size that should do the trick."

Ruth Gundersonnon spoke," Our group knows of many places on our Planet, where there is abundant heat. These would be unaffected by Climate on the surface. Your people could assist us with the equipment needed to tap this energy source, since we do not have that expertise. We do not have the technology to reach these areas; we will need your help there as well".

Joe Jonesnon spoke" There are many places on our planet where there are tremendous changes in the tides. These are created artificially and naturally. We have experts well familiar with this type of energy use, but we lack experience to build machines that can utilize this type Of energy. We can provide the manpower though."

Cintheianon and Angelanon spoke," You have some kind of an idea of our determination. What we lack in expertise, we make up for in trying to make our planet a beautiful place. We are also involved in A great deal of research. We a wait your reply".

Dave Jones spoke first," I first want to say how glad I am to be here, I have for over 20 years yearned to put our resources to work to help another world. I have Scientists, and Technicians plus a large support staff that can fill the bill on this assignment."

Robert Paul spoke," As second in command I can mobilize our Team, and your people, as well to do the tremendous logistical and Operational aspects of this project to get it into action."

James Jameson spoke," I can mobilize our manufacturing and Production to provide all of the equipment products, and personnel that we will need. We can use our large manufacturing facilities on Earth to produce what you need. In fact the Planet Alliance will provide the protection of the large Transports, thousands of meters long to transport the goods. This should help to put your planet on the mend.

The entire support staff of the Earth delegation spoke in unison," We will do all we can to Put your planet on the mend."

The entire support staff of Ropedia spoke in unison," We are grateful for your help. We look forward to working with you on this Endeavor, this is critical to our planet!!"

Just then the group from the ship Explorer 2 came in. They were Kirk Christopher, Christopher Jones, and Barbara Jones, and Rebecca James. Kirk said, "Madam Chairwomen we have been listing to your Proceedings, and since Rowpedia is a member of the Planet Alliance, We will give you full support of our ships, and whatever else Is needed to achieve this mission."

Chris, Barbara, and Rebecca spoke," We know that you will have opposition. We will advise you on this, and bring the full weight of The Planet Alliance to bear to repulse the attacks. We will also

Bring the full weight of the other members of the Planet Alliance to

bring pressure to bear on those responsible. They hide and cowl while

Millions on your world suffer from this catastrophe. We will surly put a

" kink in their corset." They raised their arms, and put their hands

together as they said this. Rebecca James said nothing, she was a Rep

For the Space Service, an Admiral who was overseeing this project for

The Space Service, and she would report to them.

   Cintheianon spoke," Thank you all. Let's adjourn to plot our strategy,

and mobilize this arsenal of help. Meeting adjourned!" With that the

meeting ended. The people went back to their quarters.

****

   Meanwhile, on another part of the planet, a different kind of

meeting   was taking place. It was a meeting place were thick

plush carpets, very thick, and soft chairs, a highly polished table and

2 chandeliers and heavy wood paneling. It was a round table with fine

Wood. It was where the big 3 major polluting groups had their many

meetings! They were the following: Joe Josephnon of Unconquerable Oil

Bob Bobbinsnon, Unbeatable Oil, and Dave Davis of Omnipotent, The

Major car companies Jack Jackson of Unbeatable Motors, Jeff

Jeffersonon of Dominate Motors, Ray raymonsnon of absolute Motors.

Last but not least Mary Martin of Unbeatable Insurance, Valerie

Valensnon of Dominate Insurance, and Rebecca Robbinsnon of the

Necessity Insurance Company. All of these men and women were in the prime of life. They were as ruthless in business as well as in personality. They would do whatever it took to get ahead. This usually dealt with the way they treated their employees, their competition, and whoever stood in their way. They, and their Minions, up till now, had an iron grip on the people of Ropedia.

Mary Martins and Dave Davis head of this "unholy Alliance" called the meeting to order. Dave and Mary spoke, "We have called this meeting, because our daughters have been stirring up the pot against Us, however, this time we have real trouble. They have called the Planet Alliance, plus a delegation From Earth to stop us. If they can turn public opinion, we are in real trouble."

Valerie and Rebecca spoke," Don't worry about that Dave, we'll make public opinion. We will stop them cold. We have the backing of People from our planet, as well as many others from other planets. They can do a great deal to discredit them.

Ray and Jack spoke," Can't you put a collar on those daughters of Yours!! We will Have to take drastic steps if necessary, but short of death though." Dave and Mary were clearly annoyed at this remark but said nothing. They tried to make them think they had no control Over their daughters.

Bob and Jeff spoke," We could put a shill in, a spy, to infiltrate, and discover their Plans. We could put a hit man on them and kidnap them, and isolate them. We maybe could discredit their group in some way."

Dave and Mary spoke," Our daughters are too smart for that, and besides they have several support people who would carry on. We

have to hit them where they live. We will not give up our huge profits,

And absolute power we have over the people."

Valerie and Rebecca spoke," We have sold insurance at sky high

prices, to drain every penny of their reserves, all of the other people

on the planet as well. The public is gullible they will buy anything,

even if it doesn't help them. We can throw in sex as well as bribes,

even if it doesn't help them. We do this by appeasement, and phony

Solicitude and service. We will employ any tactic to stop your daughters,

and their group."

Ray and Jack spoke," We have done everything possible to get the

Public to buy our products. We have destroyed the support and

and reliability of the Public Transit, mostly Buses and Trains, also

Decimated the Pod Systems so the public is totally dependent on us".

Bob and Joe spoke," We have created a phony shortage of oil, and

through political and personal prejudice to stir up wars, to jack up

the prices on our products."

All of the members spoke in unison, except Dave and Mary," We will

not give up our huge profits for ourselves, and our minions, and all of

those that are on other planets as well. We will defeat the Crusade For

A Better World with a massive Radio and T.V. blitz. We will also do the

same to the People on the other planets who support them".
The members pounded their fists on the table as they said this.

"Enough!!" said Dave and Mary "We have a formidable enemy. It will

take thought and careful planning to have it come out right, but we will succeed in the end. We call this meeting adjourned."

After the meeting was over, Dave and Mary sat in the conference Room and had a private conference.

Mary spoke," Do you really think you can stop them Dave? "

Dave said, "I do not know. There is one thing I can do, and that is, to get those guys in the Congress, who we pay enormous fees to, to pass Legislation to stop taxes, rules and regulations, and anything else that interferes with our profits."

Mary said, "Ash… yes, I'm sure they can come up with something, at least long enough to figure something out.

Just then Valerie and Rebecca came in and said, "We have a plan to discredit your daughters. Most of the people in their group, and the General public do not know of their background. I'm sure some of those Liberals would turn on them like a pack of dogs if they knew."

Dave and Mary said," We could try that, but I'm dubious of the result. They are too loyal to them, but the public might. The best solution for now is legislation in our favor for now."

The meeting ended, and they departed. Dave and Mary were confident they could find a solution.

Chapter 4

The progress of the Crusade for a Better World was gaining steam. With the Planet Alliance's help materials were pouring in. Captain Kirk

Christopher was contacting other planets, and getting their help. This

was in the form of ships, personnel, and expertise. Robert Paul and his

Wife Betty were helping the Engineers and Technicians to install the

Machines and other devices, and make them operational. The men

with Mike Mickelson of Unbeatable Wind Power were bringing

their products to Rowpedia by offering large discounts on their

Products; which were electric cars and wind power, and replacing the

Fossil Fuel types. Ray Robinson was providing assistance to get the solar

Cells operational. The people of Rowpedia purchased these products

at a huge discount and abandoning the oil and gas. John Johnson was

providing pods and tunnels for commuting to and from work and

Saving the cars for emergencies. The combination of all of these

Activities were having effect on the climate. The climate was showing

signs of beginning recovery. This was over a period of several

Months. This was planet wide so the effects quadrupled. Cintheianon

and Angelanon were astatic with joy! , with the help of Captain Kirk

Christopher, and the other ships in and around Rowpedia they had

stopped the sabotage and destruction of some of the cargo ships.

The reason for this was the ships of the Big 3 attacking the cargo ships,

did not want any part of The Planet Alliance.

Cintheianon and Angelanon had a meeting with Captain Kirk and

Crew. Cintheianon spoke," I'm indebted to you captain for stopping the

attacks, even though, they were few and far between.

Kirk Christopher spoke," Yes. They were pretty much a paper tiger

since we brought our ships To bear, and made mincemeat of them. I do not think they will bother you anymore."

Christopher Jones spoke," Yes. We have been able to deter them, in fact, even more ships from Other Planets in the Alliance have been providing materials, and expertise to stop the attacks. We can count on them to do this in the future as well."

Barbara Jones and Rebecca James said," We have applied our staff of Engineers and Technicians To get all alternate power systems going. The members of our Security Staff, and those of Rowpedia have prevented acts of sabotage".

Cintheianon and Angelanon spoke," Thank you for all of your help. We have seen progress here on the planet on this matter, and the Climate as well. We will keep you posted of same."

Meanwhile… on another part of the planet, "the unholy alliance" was Plotting a counter strike.

****

Dave Davis and Mary Martins were in the same conference Room as before. They had brought in McConnell Lorbus and Dianna Williamsnon and Bayner Carmel and Susan Samsonnon who were Their special assistants. Both of these women were very attractive, slim of build, long curly hair they both had very dark Green Eyes. They both had on Spike Heels, and Purplish-Red Lipstick. They both had on a hint of Cologne, and extremely pretty faces. They had on pantsuits for the meeting.   When doing the bidding of their leaders could be devastating to any opponent when they applied their feminine charms.

Mary called the meeting to order, and said," I called this meeting because I am going to pull out all the stops. I want you two women to do whatever it takes to trap these men, in compromising positions. To weaken their movement. To anger those bleeding hearts, to abandon them. We can turn the public from their movement."

Dianna and Susan spoke as one, "Don't worry about that Dave, we have destroyed the reliability of many an opponent, both here and abroad." They smiled as they said this.

McConnell Lorbus spoke," I can verify that! They might be the power behind the throne. The dealings Of politics is won in this way."

Bayner Carmel spoke," We can impose new subsidies to protect your huge profits. These women can truly do their job dealing with the men of power, because they understand the weakness of those that Control it." His laughter filled the chamber.

Bayner Carmel spoke," We can go through the charade of reform, and really sock it to them on tax levies on their equipment, to cripple their efforts". He laughed after this dissertation.

McConnell Lorbus spoke," We can reduce their progress on this Alternate Energy by regulating them to death. We can employ these Women to seduce the men and reduce their credibility, and their work by being involved with these women." He laughed out loud after this discourse. His laughter filled the chamber.

Dianna and Susan spoke," These men of Earth are week, like men everywhere. We can charm these men, with our feminine charms, and seduce them. We know men, and there are vain, foolish and naïve about the ways of Women!"

Dave Davis spoke," I believe ladies and gentlemen the operation should work very well. I shall meet with the associates of the Big 3, and if we find any more ideas we will let you know. We thank you for your help on this matter". We thank you for your help".

Mary spoke," I know this will not work on the men of Rowpedia. They are too dedicated to the Cause, my daughter has seen to that. The men of Earth….. you might have a chance with them, as you have said. I believe this because you two women are notorious in four Star Systems!"

Susan and Dianna spoke," We are up to the task. We can deliver the goods. We are confident our charms will win out. We have helped our Bosses many times by discrediting obstinate leaders in the past, and can do so again."

Dave and Mary smiled, the meeting was over.

****

Cintheianon and Angelanon were coordinating the attack, and doing other activities. They had the staff of Crusade for a Better World in the Field, alongside working with the people of Earth, to install and operate the equipment. This was particularly true with the tidal and Geothermal Equipment. Joe Jonesnon had stated that there were many places on the Planet, for excellent results. The reason for this;

large tidal changes of 20 to 25 meters. In fact Mike Michelson of

Unbeatable Electric and Tides, Joe Jamesnon's colleague had shown

how to Install, and operate the machines to really make them shine

The massive technical expertise of the Engineers and Technicians was

indispensable. The Geothermal, which almost unheard of on Rowpedia,

was brought to life by the efforts of Ruth Gundersonnon's colleague

Dave Jones of Alternate Energy! This had been his specialty in college.

He was able to fulfill a lifelong dream. Dave Jones was ecstatic with

joy at this prospect. He was like a kid in a candy store.

   Cathie Cuttlernon, who was a specialist in manufacturing and

Production had accelerated the manufacture of cars on Rowpedia

and through special contacts on other Alliance planets were able to

Quadruple production of solar powered cars. For enter-city travel she

worked closely with John Johnson of United Pod to build an efficient

Pod System on Rowpedia so cars would not be needed as often.

This would also help Relieve the congestion, of said vehicles.

   Pat Smithnon had help from Engineers and Technicians of Earth and

Rowpedia to get large wind farms built and operational. There

were many places on the Deserts and near the Oceans particularly

filled the bill for this. All of this activity, 5 years had gone by. The

Planet of Rowpedia was continuing to improve. The fact that this was

Planet wide the results were quadrupled.

* * * * * * * * * * * * * * * * * *

The activities of the big 3 were equal strong in opposition. The Women Dianna Williamsnon and Susan Sampsonnon had been completely squelched. The men of Earth looked at the women and what they were trying to do, and just laughed. The women were furious and their feelings were hurt, and also their pride, and they withdrew. The encounters in space, in the form of attacks grew fewer. The two Cruisers assigned to protect the ships, made mincemeat of the opposition. Captain Kirk Christopher even left, and returned to other Scientific endeavors. Bob Bobbinsnon of unbeatable Oil, and Joe Josephnon of Unconquerable Oil and all the rest of the companies where becoming desperate.

They were being hurt, where it mattered mostly, in the pocket book. They intensified their attacks on the ships and facilities, but to no avail. They obtained support from Congress for heavier tax levies on their Activities, but the ball was rolling now. The people Planet-wide were paying for the new energy sources, because they wanted them. The oil Companies Withdrew.

Ray Raymonsnon of Absolute Motors, Jeff Jeffersonon of Unbeatable Motors, and Jack Jacksonon were fighting a losing battle. The people were buying the electric cars, and converting the old ones. The people that lost their jobs at the big three were eagerly taken at the electric car company.

The men who were head of the big 3 were furious about this. They Got massive cuts in road building approved, however, the people Poured into the show rooms to get the cars anyway. This had done

Little to achieve the effect they wanted.

The insurance companies of Unbeatable head Mary Martin, Valerie Valensnon Dominate and Rebecca Robbinsnon of Necessity found that the jig was up. Not only were people not buying the phony and useless Policy's that provided no real service; or punished people with higher rates if they did, turned the tables on them. The people grew weary of the phony solicitude of the companies, that provided no real service. They banded together in co-ops and bought group planes through the Energy companies they worked for! The big 3 grew desperate. Dave Davis and Mary Martin called a meeting in the same pretentiously plush Conference room where they had all the other meetings.

Dave Davis said," ladies and gentlemen we have a grave crisis indeed. Not only have our efforts failed, but our profits have plummeted! The supporters here on Rowpedia as well as other Worlds have had a steady, and relentless pouring in of criticisms; and complaining that their losing money by the bushel!"

Mary Martin spoke," We have done everything we can think of including: taxes, bribes, sex, and sabotage and still we cannot turn the Tide of relentless abandoning of our products, and going over to our competition! Our profits continue to plummet, and our workers are going over to the competition !"

The Representatives of the other companies in the big 3 said in unison," We will not give up our huge profits. We will not yield to their Sabotage. We will not change our policies of defrauding the public and changing whatever it takes to get ahead. We will be ruthless, heartless, and ambitious in our efforts. If need be, we'll even kill to gain our ends." They pounded their fists on the conference table as they said this. They were a determined lot.

Mary Martins and Dave Davis were not impressed with this display. They came to the mike. Dave Davis said," Theatrics and fist pounding Will not stop the progress. We will have no part in killing. ! We will set up a joint Session of Congress, House and Senate to solve the matter through a Special Election".

Mary Martin said, "This will be the time to collect our dues from those guys. This will be time to get something for all of the money we give them. It will buy time to plot our next move."

Dave Davis said," All in favor say aih! All raised their hands. "All oppose" none did. "Very well, Motion carried, and meeting adjourned." After the others left, a very dejected and sad Dave Davis and Mary Martins held up their hands, and shrugged their shoulders and left!

Chapter 5

Cintheianon and Angelanon were delighted at news of speaking in front of a full Session of Congress. They were confident that they had the people on their side now. Cintheianon called a meeting of all the Key people who were: John Johnson, Mike Michelson, Ray Robinson James Jameson, Robert Paul, Betty, Robert Jenningsnon, Pat Smithnon,

Joe Jonesnon, Ruth Gundersonnon, and Cathie Carvernon. The

Key people of the Alliance, and theCusade were around the table dais

and the others in chairs around the room. A strange sight in the back

where there was no light were Dianna Williamsnon and Susan

Sampsonnon. They were unnoticed by the others and a somewhat cruel

incident had happened…… Before this….

Dianna Williamsnon and Susan Samsonnon had been at home

"licking their wounds" so to speak, when Bayner Carmel and McConnell

Lorbus paid them a visit. They said," You cheap tramps. You parade

yourselves to tease men, and us, and for the most part succeeding.

You have failed this time. We have found you have a weakness for the

Two Men in the Crusade, and have failed to seduce them, and defeat

Them. You have also leaked our strategy, and put our whole movement

in jeopardy!" They slapped them hard several times, and pushed

Them on to the sofa. Dianna and Susan were crying at this point, and

hid their faces. The real point of the anger however, was they, and

other members of the Congress had been caught in some hanky-

panky with them.

Diannna and Susan who regained some of their composure said," Is it

a crime to care for these Men? We have never had this feeling before.

We did not reveal any secrets, and the wives found out because

Someone was jealous of us, and filed the report."

Enough! , said Bayner Carmel and McConnell Lorbas. "We can have

no sentiment in our business. You both are discharged, as of right now!

We have no further use for you; you can run off with these

Men as far as we are concerned."

Dianna and Susan were crying at this when Bayner Carmel and McConnell Lorbus slapped them again and said," We will deal with these Liberals in our own way. I should have known that this plan of Seduction would fail" They pushed the two women on to the couch, and left in anger.

McConnell Lorbas and Bayner Carmel had similar backgrounds. They both went to separate schools but meet in college. They both were active in sports; in fact they were 'big man' on campus, because of This. McConnell lorbas was golf, and Bayner Carmel was baseball. They were extremely popular with the Ladies, and never knew a lonely time. A lot of people thought highly of them, and felt they were destined for Greatness, and They were destined for Government Service. They excelled in Government, and Law in College. They started out as young Lawyers, in their respective House, and moved up rapidly to head of each. At this time they were still unstained. Then they turned bad through backroom deals, corrupt politics, and a general ill-informed Electorate led to their current positions. They were constantly at Swords point with the Crusade, And with Cintheianon and Angelanon, and anyone else who stood in their way.

After a few minutes the women (sitting on the couch) decided to go to the meeting, of the Crusade for a Better World, and see what they could do to help them. Upon arrival at the meeting, they said in low tones, "I envy their dedication. They haven't even noticed that we are Here. They figure that we are useless as well". The meeting proceeded as Cintheianon and Angelanon were about to speak.

Cintheianon and Angelanon said, "We have our key people here I would like to call this meeting To order".

We have your people, plus the people from Earth here, and are here to make arrangements to Speak before the Congress', said Cinthianon

Robert Paul said, "We have my wife Betty here as well as the following: John Johnson from United Pod Systems, Roy Robinson of Full Power Solar, Mike Michelson of Unbeatable Electric We have the Key people from the Crusade as well, they are the following; Robert Jenningsnon from Freedom Solar, Joe Jonesnon from Eternal Tides and Ruth Gundersonnon from Dependable Geothermal. We have in addition to that, we have Cathie cuttlernon chief Manufacturing expert from Rowpedia".

James Jameson spoke," I'm from manufacturing and production, and I'm here too." Every one Laughed at this.

The group from the Crusade spoke in unison and said," We are delighted with the progress, and look forward to dealing directly with the Congress".

Cintheianon and Angelanon said,' we have a meeting with them, immediately after we are done here. We are very hope full for this Election. We should be able to defeat the 'big 3' polluters after this meeting with the Congress and the Election".

Just then… Bayner Carmel and McConnell Lorbus rudely and noisily entered the room. They said," We look forward to this election and Campagne as well. We will have a chance to really show you up.

In fact I can inform you that we have traitors in our midst. They will

betray you as they betrayed us".

He pointed to the two women in the back of the hall. They said," You

have betrayed, and disgraced us. You have betrayed our cause; we

will put an end to this." Bayner Carmel and McConnell lorbus

Slapped the faces of the two women.

Robert Jenningsnon and Joe Jonesnon got up from the table, and

grabbed the two Congressmen, and pushed them against the wall.

They said," That's no way to treat women, even if they decided

Not to support you. We are more determined than ever to defeat you."

Angelanon spoke," High noon tomorrow gentlemen, to start the

Campaigne and then the Election. Now get out of here!"

Bayner Carmel, and McConnell Lorbus left the hall. Robert

Jenningsnon and Joe Jamesnon came over and put their arms around

The two women. They came to the front of the room and the dais,

and introduced them to the group.

Dianna and Susan said," We are familiar with the halls of power.

We know where the bodies are buried of the big 3. We know nothing

about Pollution or the other fields involved, but we understand power

and the men who control it. We offer our help. We can speak out

against the big 3, this will help in this campagne we need the men

Of your group though".

Cintheianon and Angelanon spoke," We are grateful for the help, we

are glad to protect you. I Believe that Joe Jonesnon and Robert

Jenningsnon will do this".

Because of the disturbance of the Congressman, the meeting was

over, and the people left. Joe and Robert came over to the two women

to comfort them. Joe put his big arms around Susan's neck, and

Caressed it and her cheek, and gave her a big kiss full on the lips. Then

gave her a long hug. Robert went over to Dianna and did the same He

had big arms too.

Joe and Robert spoke," When you tried to seduce us, and persuade

us with sexual favors to abandon the Crusade, we doubted you. We

are happy to protect you against attack."

Diannna and Susan spoke," We believe from the beginning that we

had some feeling for you. We just couldn't carry through with our

Mission. We told the Congressman of our plan to leave; and that's when

they slapped us, and abused us, and pushed us around."

Joe and Robert caressed the women on the check, and gave them a

full kiss on the lips and said, "We love you very much, and will always

protect you." They each took their hand, and walked out of the hall.

Chapter 6

The campaigns began in earnest. The speeches were numerous, and

emotional. Before a joint secession of Congress, Cinthianon and

Angelanon gave emotional poignant speeches. At the close

They said," We know that we have been enemies; however, since the

Tide has turned; we can work together for a better world." This was met by almost total silence. No response from the Congress. They, Angelanon and cintheianon were deeply disappointed.

The big 3 realizing that the jig was up, made many speeches, one of the most poignant was before The Council of Planets, Dave Davis and Mary Martins spoke, and in conclusion said," We have had a cozy Relationship. This has been profitable for both sides. This relationship will end, if the Alliance is Not defeated. The food will be taken from Millions of workers, who will lose their jobs over this. There will be many children go hungry!" This hit a very powerful cord with the people.

The Governing Council, who was responsible for all civic matters on Rowpedia on all the Continents on the Planet; which had said very little up to this point stated in unison," We have faith in the Planet Alliance and the Crusade For A Better World. They have worked tirelessly, For a better planet. We have seen in the past 10 year's tremendous improvement. We are behind your group 100 %." This was spoken To Cinthianon and Angelanon, along with the following; the entire Staff of the Crusade, all of the Representatives of the Planet alliance headed by Dave Davis, Robert Paul, his wife Betty, this included all of the heads of departments from Earth and their support staff. They applauded after this was said.

Dianna Williamsnon and Susan Sampsonnon spoke in a huge Stadium filled with people, and on a planet wide hook-up; told of the fraud, deceit, ruthlessness, and selfishness of the big 3. They closed with the following: "We have served these Men to help them in

Operations And sexual doings, and all other activities. We know where all the Bodies are buried. When we refused this time, they beat us up, and abused us!" This particularly hit a strong cord, due to the fact of the Non-violence Directive Towards Women on Rowpedia.

All of the executives of the big 3 spoke in unison at a large park with another planet wide hook-up said in conclusion, "We can show many instances of the people of the Alliance had fraud. We can show where they used the same tactics, they accuse us of using. We have had a cozy relationship, and One that has been profitable for both. This will come to an end if they, the Alliance, is not defeated. This has to be decisive and complete! "This particular claim could never be proven however, and this was one more nail in the coffin for the big 3. They were very angry about that.

****

The election was held to determine the fate of the planet. There were some disruptions, voting snafus, and other confusion, but by and large it was relatively free of violence. The returns were covered by all the media. The turnout was close to 95%. It could have been been higher, but there were some logistical problems. The over whelming majority of the people voted to continue the program Of the Crusade, and the Planet Alliance. Cintheianon and Angelanon were estatic with joy! There was a victory party, however it was Robert Paul and his wife Betty who spoke, "We are grateful for this Victory but, it is not a victory for us, but for your planet. Dave and

Betty, and I have realized a life long dream by helping you. Your

Leaders Cintheianon and Angelanon who never lost hope, or faith

In this project deserve the most praise! They are the ones mainly

responsible for the salvation of your world!"

The people cheered and cheered, but then grew silent, realizing

the truth of the words spoken. Then they departed. The cheering was

accompanied by Dave Jones, and Robert Paul hugging Cinthianon and

Angelanon, and shaking their hands. Betty did the same thing, and

then departed.

Dave Davis and Mary Martin, who were in the same plush meeting,

Room tried to console the group. The Big 3 executives were dejected,

and were determined to make the most of the situation. They

Realized that the few who were left, would vote them out, and the

Companies would be absorbed into the Crusade. The CEO'S were so

hard core however, that instead of working in the Crusade, left

The planet as a group. They were unbent, and unrepentant in their

Deeds, and were consigned to their fate.

Bayner Carmel and McConnell Lorbas were disgraced, forced to

resign due to improprieties, misconduct, corruption, and just immoral

behavior towards the women Dianna, and Susan. The other Congress

Members who were involved, the same happened to them.

A special council was set up to have a comition to appoint a new

Congress. An election was held, and 80 % of the Congress was

replaced! Bayner Carmel, and McConnell Lorbus were exiled to a Moon

of Rowpedia. The others involved meet the same fate. The Moon had a harsh climate, where survival would be paramount. There was no chance for the Group to have a chance for corruption.

Cintheianon and Angelanon were appointed to be head of each House of Congress to help guide them in the future. The other people for The Crusade for a Better World continued as heads of their respective companies and to work with the Congress. The people who were support Engineers, and Technicians, and Support staff continued in their respective jobs.

Dianna Williamsnon who returned to her wearing nice clothes, Makeup, a hint of cologne and with her spike heels was married to Robert Jenningsnon after a short courtship. They had a relatively Quiet life after that. He continued as head of Freedom Solar. Susan Sampsonnon did the same with Joe Jonesnon. and he continued as the head of Eternal Tides. They often went to plays, and concerts With Dianna and Robert on Rowpedia.

The people who were in charge of the various agencies, who had come from Earth with Dave and Robert and Betty returned to Earth to resume their lives Robert Paul and his wife Betty continued to work for Alternate Energy. They went to Plays and Concerts on Earth with Their friends Dave and Clairisa. All and all they had been gone for a little more than 10 years, and Clarisa, the faithful wife was there when Dave returned to Earth. She was astatic with joy. She caressed his neck with her fingers. She threw her arms around Dave and gave him a long kiss, this lasted for several minutes. After this she said," I'm very proud of you, and what you have done. I'm sorry that there are no big crowds here to greet you and your associates. I heard there were latterly

Millions of them on Rowpedia."

Dave smiled at this and said," That's alright, I never liked crowds

anyway! What was it you wanted to talk over with me, I'm curious

about that?" He looked at her somewhat quizzically.

Clairisa said, "I want to have a child, a product of our love. I want

Something to remind me of you and how together we can nurtcher

that child to maturity."

Dave was very pleased about this and said," Well I'll be! I couldn't

be happier, and more delighted to have a child, it has to be a boy of

Course! (He winked)

Clarisa said, "I would like to have a girl, but a boy would be ok as

Well. We could have girl talk during the time you are gone. I always

wanted a daughter; I came from a family of all boys".

Dave said," But wouldn't a boy be a reminder, when I get old, a

younger version of myself."

Clarisa smiled and said," that's silly dear, it doesn't matter. "She gave

Him a long kiss on the lips. It would seem the fates would have it, they

Both were satisfied, they had twins!!

THE END